THE WORLD OF OCEAN ANIMALS

SEA STARS

by Adeline J. Zimmerman

Ideas for Parents and Teachers

Pogo Books let children practice reading informational text while introducing them to nonfiction features such as headings, labels, sidebars, maps, and diagrams, as well as a table of contents, glossary, and index.

Carefully leveled text with a strong photo match offers early fluent readers the support they need to succeed.

Before Reading

- "Walk" through the book and point out the various nonfiction features. Ask the student what purpose each feature serves.
- Look at the glossary together. Read and discuss the words.

Read the Book

- Have the child read the book independently.
- Invite him or her to list questions that arise from reading.

After Reading

- Discuss the child's questions. Talk about how he or she might find answers to those questions.
- Prompt the child to think more. Ask: Have you ever seen a sea star? If so, where was it? If not, would you like to?

Pogo Books are published by Jump!
5357 Penn Avenue South
Minneapolis, MN 55419
www.jumplibrary.com

Library of Congress Cataloging-in-Publication Data

Names: Zimmerman, Adeline J., author.
Title: Sea stars / Adeline J. Zimmerman.
Description: Minneapolis: Jump!, Inc., 2022.
Series: The world of ocean animals
Includes index. | Audience: Ages 7-10
Identifiers: LCCN 2021022670 (print)
LCCN 2021022671 (ebook)
ISBN 9781636902944 (hardcover)
ISBN 9781636902951 (paperback)
ISBN 9781636902968 (ebook)
Subjects: LCSH: Starfishes–Juvenile literature.
Classification: LCC QL384.A8 Z57 2022 (print)
LCC QL384.A8 (ebook) | DDC 593.9/3–dc23
LC record available at https://lccn.loc.gov/2021022670
LC ebook record available at https://lccn.loc.gov/2021022671

Editor: Jenna Gleisner
Designer: Michelle Sonnek

Photo Credits: HHelene/Shutterstock, cover; Tiago Sa Brito/Shutterstock, 1; Ethan Daniels/Shutterstock, 3, 18-19tr; Adam Ke/Shutterstock, 4; David Fleetham/Alamy, 5; Nick Greaves/Shutterstock, 6-7; Richard Whitcombe/Shutterstock, 8-9, 18-19tl; DarrenMower/iStock, 10-11; Andrew J. Martinez/Science Source, 12, 13; Dominic Robinson/Alamy, 14-15; Ryan McGurl/Shutterstock, 16; Tulen/Shutterstock, 17; alonanola/Shutterstock, 18-19bl; Gerald Corsi/iStock, 18-19br; Lawrence Naylor/Science Source, 20-21; wcpmedia/Shutterstock, 23.

Printed in the United States of America at Corporate Graphics in North Mankato, Minnesota.

TABLE OF CONTENTS

CHAPTER 1

STARS OF THE SEA

What sea creature can regrow lost arms? This ocean animal is a sea star!

There are around 2,000 sea star **species**. Most have five arms, like a star. This is how they got their name. But some can have up to 40!

Sea stars are often called starfish. But they are not fish. They do not have fins, gills, or scales. They are **echinoderms**. Hard, bumpy skin covers their topsides.

Tiny eyespots are on the tips of each arm. These aren't like eyeballs. But they can **sense** light and shapes. This helps sea stars find food.

DID YOU KNOW?

Sea stars don't have brains! They don't have blood either.

Sea stars live in many **habitats**. These include **tide pools**, rocky shores, and **coral reefs**. Some live in warm, shallow water. Others live in very cold, deep water.

TAKE A LOOK!

There are no freshwater sea stars. They only live in oceans. Take a look!

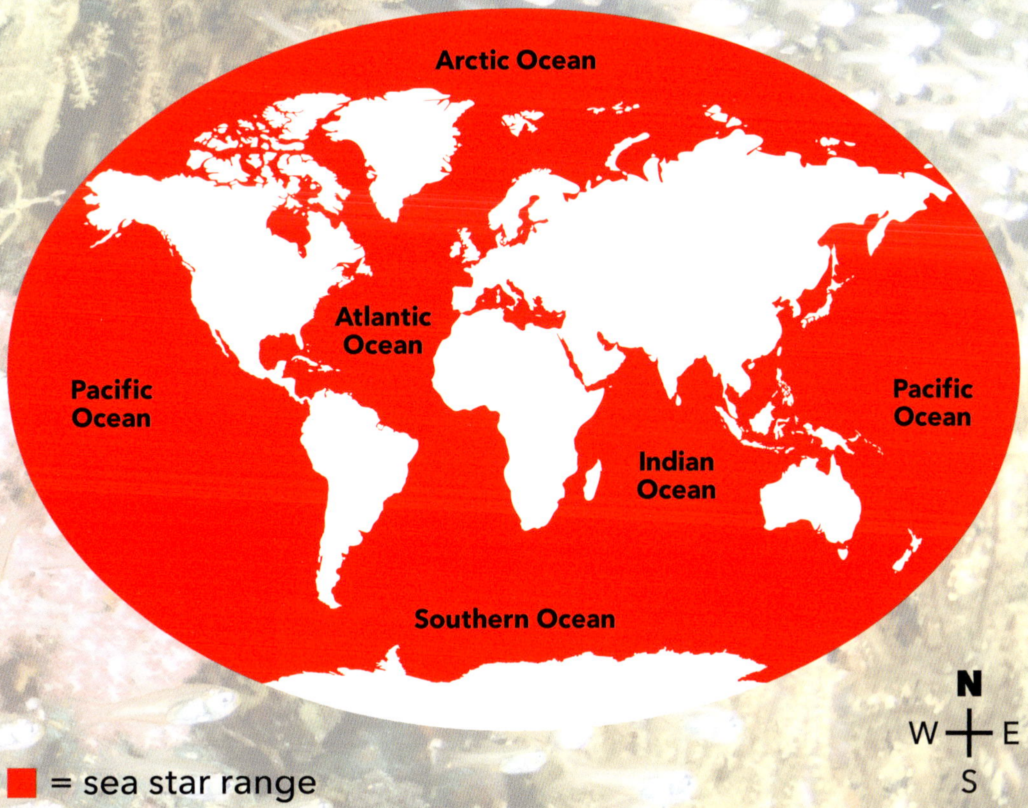

Sea stars are **invertebrates**. They do not have any bones. On their undersides are rows of tiny tube feet. The tube feet act as **suction cups**. Sea stars use these to move along the ocean floor. They also help them stick to surfaces, like rocks.

TAKE A LOOK!

What are a sea star's body parts called? Take a look!

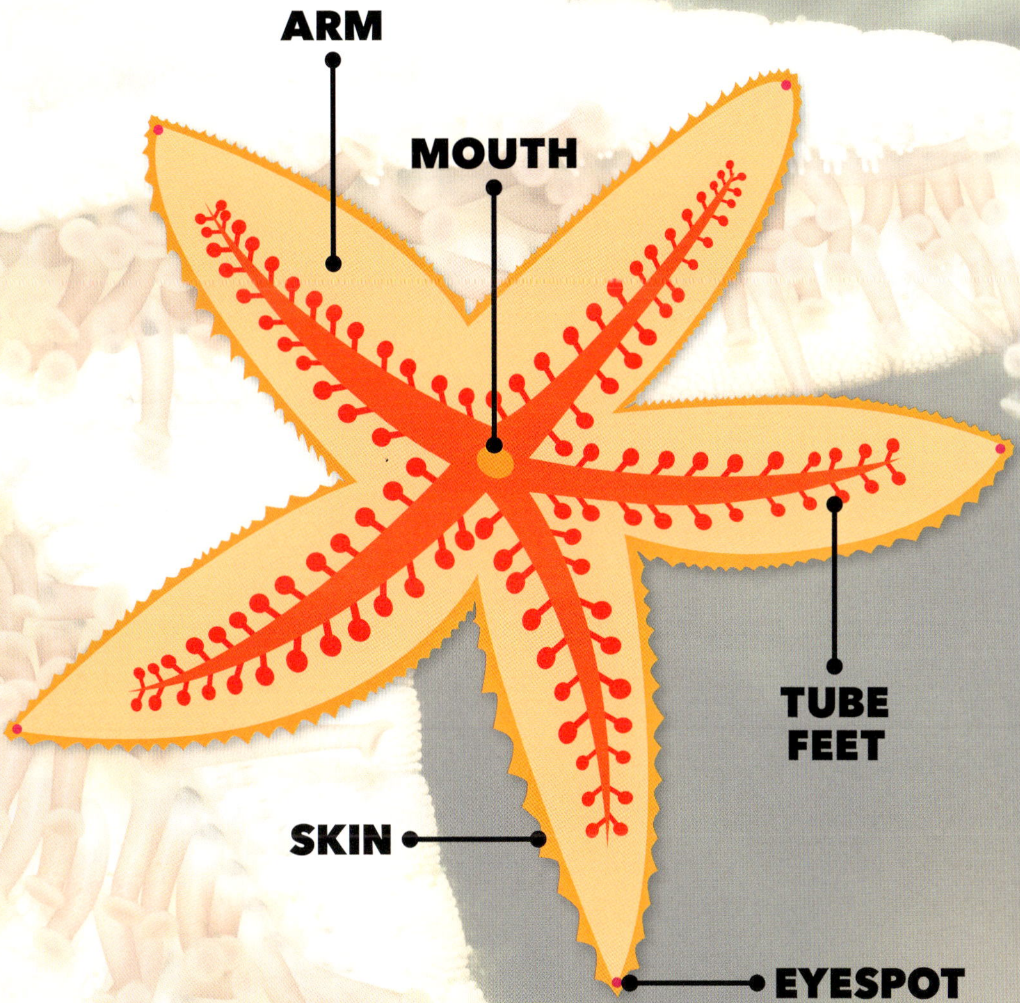

CHAPTER 2

PREDATOR AND PREY

Tube feet help sea stars eat. Sea stars are **carnivores**. They eat clams, oysters, and mussels. They use their tube feet to pry open the shells of their **prey**.

Then, things get interesting. The sea star's stomach comes out of its mouth. It oozes inside its prey's shell. **Enzymes** from the stomach **digest** the meal. When it's done, the stomach goes back into the sea star's body.

When it comes to a sea star's **predators**, sea otters, fish, and birds are just a few. Hard skin helps protect sea stars. But larger predators can swallow sea stars whole. Others may take a bite. But don't worry! Sea stars can regrow lost arms.

DID YOU KNOW?

Growing back an arm doesn't happen quickly. It can take up to one year!

CHAPTER 3

SUPERSTARS

Sea stars come in many colors and **textures**. Some, like the royal sea star, are bright. Bright colors tell predators to stay away.

For others, color is **camouflage**. The southern sand star is tan and brown. It blends in with sand. It is hard for predators to see.

crown-of-thorns sea star

chocolate chip sea star

granulated sea star

morning sun star

The crown-of-thorns sea star is prickly. **Spines** help protect it. Chocolate chip sea stars have black points. Its points and bright orange color warn predators.

The granulated sea star is also called the cushion star. Why? It is puffy! The morning sun star is bright orange. It looks like a sun. It has up to 16 arms!

At more than three feet (1.0 meter) across, the sunflower sea star is the largest. It is also the fastest! It uses up to 24 arms to move almost 10 feet (3.0 m) per minute.

Unique sea stars live in every ocean. They can be found on beaches around the world. Have you seen one?

DID YOU KNOW?

The smallest sea star could fit on your fingernail! The paddle-spined sea star is less than one inch (2.5 centimeters) across. Its tube feet are so small you'd need a microscope to see them move!

sunflower
sea star

ACTIVITIES & TOOLS

TRY THIS!

RATING ROUGH TEXTURES

Sea stars come in many different textures. Their rough skin helps protect them from predators. Observe and rate different textures with this fun activity!

What You Need:

- different textured objects, such as carpet, tile floor, brick, sidewalk, tree bark, grass, and rocks or pebbles
- notebook or sheet of paper
- pencil or pen

1. **Find 10 objects or surfaces with different textures. Write the name of each object in a notebook or on a sheet of paper.**
2. **Slowly run your hand over each object. Notice how it feels. Is it smooth? Is it rough? Are there any surfaces that are bumpy, like a sea star's skin?**
3. **Rate the roughness of each object. Which is the smoothest or softest? Number that object #1. Number them from smoothest to roughest so that the roughest is #10. What did you notice about the objects? Were you surprised by your findings?**

GLOSSARY

camouflage: A disguise or natural coloring that allows animals to hide by making them look like their surroundings.

carnivores: Animals that eat meat.

coral reefs: Long lines of coral that lie in warm, shallow waters.

digest: To break down food so that it can be absorbed and used by the body.

echinoderms: Sea creatures that have similar arms and rough or pointy skin.

enzymes: Proteins produced by an animal or plant that cause chemical reactions to occur inside.

habitats: The places where an animal or plant is usually found.

invertebrates: Animals that do not have backbones.

predators: Animals that hunt other animals for food.

prey: Animals that are hunted by other animals for food.

sense: To feel or become aware of something.

species: One of the groups into which similar animals and plants are divided.

spines: Hard, sharp, pointed growths on some animals and plants.

suction cups: Round, shallow cups that press against and stick to a surface.

textures: The ways things feel.

tide pools: Pools of water left behind when the ocean tide is low.

INDEX

TO LEARN MORE

Finding more information is as easy as 1, 2, 3.

1. Go to www.factsurfer.com
2. Enter "seastars" into the search box.
3. Choose your book to see a list of websites.

FACT SURFER